Am I a Monkey?

Am I a Monkey?

Six Big Questions about Evolution

❦ FRANCISCO J. AYALA ❦

The Johns Hopkins University Press
Baltimore

© 2010 The Johns Hopkins University Press
All rights reserved. Published 2010
Printed in the United States of America on acid-free paper
9 8 7 6 5 4 3 2 1

The Johns Hopkins University Press
2715 North Charles Street
Baltimore, Maryland 21218-4363
www.press.jhu.edu

Library of Congress Cataloging-in-Publication Data

Ayala, Francisco José, 1934–
    Am I a monkey? : six big questions about evolution / Francisco J. Ayala.
        p.   cm.
    Includes bibliographical references.
    ISBN-13: 978-0-8018-9754-2 (hardcover : alk. paper)
    ISBN-10: 0-8018-9754-8 (hardcover : alk. paper)
    1. Evolution (Biology) 2. Life—Origin. I. Title.
    QH366.2.A968 2010
    576.8—dc22   2010001245

A catalog record for this book is available from the British Library.

Special discounts are available for bulk purchases of this book.
For more information, please contact Special Sales at 410-516-6936 or
specialsales@press.jhu.edu.

The Johns Hopkins University Press uses environmentally friendly book
materials, including recycled text paper that is composed of at least 30 percent
post-consumer waste, whenever possible. All of our book papers are acid-free,
and our jackets and covers are printed on paper with recycled content.

To Hana, with love

Contents

ACKNOWLEDGMENTS   ix

INTRODUCTION   xi

1   Am I a Monkey?   3

2   Why Is Evolution a Theory?   17

3   What Is DNA?   29

4   Do All Scientists Accept Evolution?   49

5   How Did Life Begin?   61

6   Can One Believe in Evolution *and* God?   73

BIBLIOGRAPHIC NOTE   85

Acknowledgments

I am extremely grateful to Vincent J. Burke, senior editor at the Johns Hopkins University Press. This book was conceived by him, and he invited me to write it. It has been a pleasure to work with Vince, who has consistently provided good advice, always graciously. Others at the JHU Press have also been gracious and helpful, particularly Julie McCarthy, managing editor; Jennifer E. Malat, acquisitions assistant; and George Roupe, a superb copy editor. My heartfelt thanks to all. They made the experience of writing this short book most enjoyable. My debt to Denise Chilcote for preparing the manuscript is very extensive, and so is my gratitude. As my executive assistant for twenty-three years, she has never failed to do whatever needs to be done as well as it could be done. I remain amazed at the enormous amount of work that she is able to accomplish. I have tremendously benefited from her dedication and perfectionism.

Introduction

Darwin completed the scientific revolution by extending to the living world the notion that the workings of the universe can be explained through natural laws. He provided compelling evidence of the evolution of organisms. More important yet is that Darwin discovered natural selection, the process that explains the "design" of organisms. The adaptations and diversity of living beings, the origin of novel and complex species, even the origin of humankind could now be explained by an orderly process of change governed by natural laws.

The scientific revolution that took place in the sixteenth and seventeenth centuries ushered in a conception of the universe as matter in motion governed by natural laws. The discoveries of Copernicus, Galileo, Newton, and other scientists brought about a fundamental revolution, namely, a commitment to the postulate that the universe obeys immanent laws that can account for natural phenomena. But the diversity and adaptations of living beings had been left out of science.

The functional design of organisms seemed to call for a Designer, who would account for animals' having eyes to see, wings to fly, and fins to swim and plants' having chlorophyll for photosynthesis. Humankind's conception of the universe had become double edged. Scientific explanations derived from natural laws dominated the world of nonliving matter on Earth as well as in the heavens. Supernatural explanations, which depended on the unfathomable deeds of a Creator or other preternatural agencies, accounted for the origin and configuration of the living creatures—the most diversified, complex, and interesting realities in the world.

It was Darwin's genius to resolve this conceptual dichotomy. Darwin's discovery of natural selection completed the scientific revolution by drawing out for biology the notion of nature as a lawful system of matter in motion that human reason can explain without recourse to supernatural agencies. Natural selection is one of the greatest ideas anyone has ever had, and Darwin is one of the most influential scientists of all time.

The theory of biological evolution is the central organizing concept of modern biology. Evolution scientifically explains why there are so many different kinds of organisms, and it accounts for their similarities and differences. It accounts for the appearance of humans on Earth and reveals humans' biological connections with other living beings. It provides an understanding of constantly evolving bacteria, viruses, and other pathogens and enables the development of effective ways to protect ourselves against the diseases they cause. Knowledge of evolution has made possible advances in agriculture, medicine, and biotechnology.

The theory of evolution is perceived by many people, particularly but not only in the United States, as controversial. This perception is surprising to me, a geneticist and evolutionist who has dedicated his life to studying the evidence and the processes that account for evolution. It is beyond reasonable doubt that organisms, including humans, have evolved from ancestors that were very different from them. The evolution of organisms is accepted by scientists with the same degree of confidence as they accept other well-confirmed scientific theories, such as the revolution of the earth around the sun, the expansion of galaxies, the atomic theory, or the genetic theory of biological inheritance.

Am I a Monkey? Six Big Questions about Evolution seeks to explain some central tenets of the theory of evolution by answering questions that arise in the minds of people who are only vaguely familiar with the notion of evolution. Two points need to be made at the outset. The first is that science is a wondrously successful way of knowing the world, but it is not the only way. Knowledge also derives from other sources, such as common-sense experience, imaginative literature, music and artistic experience, philosophical reflection, and, for people of faith, religion and revelation.

The second point, to which I dedicate only the last chapter of the book, is related to the first. Science and religion need not be in contradiction. Indeed, if they are properly understood, they cannot be in contradiction because they concern different matters. Successful as it is, and universally encompassing as its subject is, a scientific view of the world is hopelessly incomplete. Matters of value and meaning are outside science's scope. In order to understand the purpose and meaning of life, as well as matters concerning moral and religious values, we need to look elsewhere. For many people these questions are very important, even more important than science per se.

Am I a Monkey?

1

Am I a Monkey?

I AM A PRIMATE. Monkeys are primates, but humans are not monkeys. Primates include monkeys, apes, and humans. Humans are more closely related by descent to apes than to monkeys. That is, the apes are our first cousins, so to speak, while the monkeys are our second or third cousins. Among apes we are most closely related to the chimps, less so to the gorillas, and even less to the orangutans. The human lineage separated from the chimp lineage about 6 or 7 Ma (million years ago). We know about these matters in three ways: by comparing living primates, including humans, with each other; by discovery and investigation of fossil remains of primates that lived in the past; and by comparing their DNA, proteins, and other molecules. DNA and proteins give us the best information about how closely related we are to each of the primates and they to each other. But to learn how human lineage changed over time as our ancestors became more and more humanlike, we have to study fossils.

Darwin's theory of evolution asserted that humans and apes share common ancestors, which were not human. His contemporaries questioned where the "missing link" was, the intermediate organism between apes and humans. Darwin published his best-known book, *The Origin of Species*, in 1859; in 1871, he published *The Descent of Man*, which extends the theory of evolution to humans; he died in 1882. Primates that were ancestors to humans after our lineage separated from the chimp lineage are called hominids (or hominins). At the time of Darwin's death, no hominid fossils ancestral to modern humans were known, although he was persuaded that they would eventually be found. The first hominid fossil was discovered in 1889 by a Dutch physician, Eugene Dubois, on the island of Java. It consisted of a femur and a small cranium. Because he was expert in human anatomy, Dubois knew that these fossils belonged to an individual with bipedal gait; the femur was very similar to the femur of a modern human. But the capacity of the small cranium was about 850 cc (cubic centimeters), which could hold a brain weighing somewhat less than 2 pounds (a pound is 454 grams, equivalent to 454 cc), while the cranium of a modern human is about 1,300 cc (with a brain of about 3 pounds). The fossil discovered by Dubois was from an individual who lived about 1.8 Ma and is now classified in the species *Homo erectus*. Our own species is called *Homo sapiens*.

The "missing link" is no longer missing. The fossil from Java was the first one, but hundreds of fossil remains belonging to hundreds of individual hominids have been discovered in the twentieth and twenty-first centuries in Africa, Asia, and Europe and continue to be discovered at an accelerated rate. These fossils have been studied and dated using radiometric and other methods. Some fossil hominids are very different from others, as well as from humans, and are classified in different species. The record of fossil hominids that lived at different times shows that several changes occurred through time in the lineage of modern humans. One change was increase in body size; another was increase in cranial capacity (and brain size). The species names are sometimes exotic, referring in some cases to the place where the fossils were found or their morphological characteristics and determined in others by the whim of the discoverers.

The oldest known fossil hominids are 6 to 7 million years old, come from Africa, and are known as *Sahelanthropus* and *Orrorin*. Their anatomy indicates that they were predominantly bipedal when on the ground, but they had very small brains. *Ardipithecus* lived about 5.5 Ma, also in Africa. Numerous fossil remains from diverse African origins are known of *Australopithecus*, a hominid that appeared about 4 Ma. *Australopithecus* had an upright human stance but a cranial capacity of about one pound, comparable to that of a gorilla or chimpanzee and about one-third that of modern humans. The skull of *Australopithecus* displayed a mixture of ape and human characteristics—a low forehead and a long, apelike face but teeth proportioned like those of humans. Other early hominids partly contemporaneous with *Australopithecus* include *Kenyanthropus* and *Paranthropus*; both had comparatively small brains, although some species of *Paranthropus* had larger bodies. *Paranthropus* represents a side branch of the hominid lineage that became extinct.

Much more similar to us are hominids classified as *Homo habilis*, the earliest species classified in the same genus as our own, *Homo*. These ancient individuals made very simple stone tools, the first hominids to do so, which is why they were given the name *habilis*, Latin for "handy" or "skilled." They had a cranial capacity of about 600 cc, greater than that of any of the earlier hominids, but less than half the brain size of modern humans. *Homo habilis* lived in tropical Africa between 2.5 and 1.5 Ma. In *Homo habilis* we can see the modest beginnings of human technology.

Homo habilis was succeeded by *Homo erectus*, which evolved in Africa somewhat before 1.8 Ma, had a cranial capacity of 800 to 1,100 cc (2 to 2.5 pounds), and made tools more advanced than those of *Homo habilis*. Two features of *Homo erectus* deserve particular attention. One is that the species persisted (with relatively small morphological changes at various times and places) for a long time, from 1.8 Ma up to nearly 400,000 years ago (ya). A second distinctive feature of *Homo erectus* hominids is that they were the first intercontinental wanderers. Shortly after their emergence in Africa, they spread to Europe and Asia, reaching as far as northern China and Indonesia (where Dubois found the first hominid fossils ever discovered) as early as 1.6–1.8 Ma, and persisted until perhaps 250,000 ya.

Two hominid species that evolved after *Homo erectus* are *Homo neanderthalensis* and *Homo sapiens*, our own. Numerous *Homo neanderthalensis* ("Neanderthal man") fossils have been found in Europe, where they first appear about 200,000 ya and became extinct 30,000 ya. The most recent fossils of *Homo neanderthalensis* were found in Spain, where they seemingly had their last abode. Neanderthals had large brains, much like ours, and bodies also similar to ours but somewhat stockier.

The evolution from *Homo erectus* to *Homo sapiens* may have started about 400,000 ya, from which time fossils have been found that are considered "archaic" forms of *Homo sapiens*. Anatomically modern humans evolved in Africa around 200,000 or 150,000 ya and eventually colonized the rest of the world, replacing other hominids. The *Homo erectus* that had earlier colonized Asia and Europe did not leave any direct descendants. (A possible exception is *Homo floresiensis*, the minuscule hominids whose fossil remains were discovered in 2004 on the Indonesian island of Flores, where they lived 12,000–18,000 ya. They may have been direct descendants of Asiatic *Homo erectus*, although the matter is still being investigated.)

The colonization of the world continents by modern *Homo sapiens* is relatively recent: Southeast Asia and the region that is now China 60,000 ya and Australasia shortly afterward; Europe only about 35,000 ya; America about 15,000 ya by colonizers from Siberia. Ethnic differentiation among human populations is therefore relatively recent, a result of divergent evolution between geographically separated populations during the past 60,000 years.

The Human Genome Project of the United States was initiated in 1989, funded through two agencies, the National Institutes of Health and the Department of Energy. (A private enterprise, Celera Genomics, started in the United States somewhat later but joined the government-sponsored project in achieving, largely independently, similar results.) The goal was to get the complete sequence of one human genome in fifteen years at an approximate cost of $3 billion, coincidentally about one dollar per DNA letter. A draft of the genome sequence was completed ahead of schedule in 2001. In 2003, the Human Genome Project was finished.

Obtaining the DNA sequence of a human genome was a great technological feat. As pointed out in chapter 3, printing the 3 billion nucleotides sequence of the DNA of a human genome would require a thousand volumes of a thousand pages each. The feat of sequencing such an enormously long genome was made possible by the development of new technologies that now make it simple, relatively speaking, to sequence genomes of that size. The DNA genomes of a few human individuals have been sequenced in the last few years, at a cost each of only about $100,000 (compared with the cost of $3 billion for the first human genome sequenced) and requiring only about one month, rather than fourteen years.

The DNA genome sequences of many other species have now been completed, in particular that of the chimpanzee, first published on September 1, 2005. Comparisons between the two genomes are being made, seeking to understand what it is at the genetic level that makes us distinctively human. It came as a surprise to many that in the genome regions shared by humans and chimpanzees, the two species are 99 percent identical. This difference appears to be very small or quite large, depending on how one chooses to look at it: 1 percent of the total seems very little, but it would amount to a difference of 30 million DNA letters out of the 3 billion in each genome.

When we compare the genomes in greater detail, it turns out that 29 percent of the enzymes and other proteins encoded by the genes are identical in both species. Out of the one hundred to several hundred amino acids that make up each protein, the 71 percent of nonidentical proteins differ between humans and chimps by only two amino acids on average. If one takes into account DNA segments found in one species but not the other, the two genomes are only about 96 percent identical. That is, a large amount of genetic material—about 3 percent, or some 90 million DNA letters—has been inserted or deleted since humans and chimps initiated their separate evolutionary ways, 6 to 7 Ma. Most of this DNA does not contain genes coding for proteins.

Comparison of the two genomes has provided insights into the rate of evolution of particular genes in the two species. One finding is that genes active in the brain have changed more in the human lineage than in the chimp lineage. On the whole, 585 genes, including genes involved in resistance to malaria and tuberculosis, have been identified as evolving faster in humans than in chimps. (Note that malaria is a much more severe disease for humans than for chimps.) There are several regions of the human genome that seem to contain beneficial genes that have rapidly evolved within the past 250,000 years. One region contains the *FOXP2* gene, involved in the evolution of speech, discussed in chapter 3.

We now know some basic features that contribute to human distinctness: the large brain and the accelerated rate of evolution of some genes, such as those involved in human speech. This knowledge is of great interest, but what we so far know advances but very little our understanding of what genetic changes make us distinctively human.

Extended comparisons of the human and chimp genomes and experimental exploration of the functions associated with significant genes will advance considerably our understanding, over the next decade or two, of what it is that makes us distinctively human. The distinctive features that make us human begin early in development, well before birth, as the linear information encoded in the genome gradually becomes expressed into a four-dimensional individual, an individual who changes in configuration as time goes by. In an important sense, the most distinctive human features are those expressed in the brain, those that account for the human mind and for human identity.

One outcome to take into account as we seek to know what makes us distinctively human and so different from other primates is that, with the advanced development of the human brain, biological evolution has transcended itself, opening up a new mode of evolution: adaptation by technological manipulation of the environment. Organisms adapt to the environment by means of natural selection, by changing their genetic constitution over the generations to suit the demands of the environment. Humans (and humans alone, at least to any significant degree) have developed the capacity to adapt to hostile environments by modifying the environments according to the needs of their genes. The discovery of fire and the fabrication of clothing and shelter have allowed humans to spread from the warm tropical and subtropical regions of the Old World, to which we are biologically adapted, to the whole earth except for the frozen wastes of Antarctica. It was not necessary for wandering humans to wait until genes providing anatomical protection against cold temperatures by means of fur or hair would evolve. Nor are we humans biding our time in expectation of wings or gills; we have conquered the air and seas with artfully designed contrivances—airplanes and ships. It is the human brain (or rather, the human mind) that has made humankind the most successful living species by most meaningful standards.

One exciting biological discipline that has made great strides within the past two decades is neurobiology. An increased commitment of financial and human resources to that field has enabled an unprecedented rate of discovery. Much has been learned about how light, sound, temperature, resistance, and chemical impressions received in our sense organs trigger the release of chemical transmitters and electric potential differences that carry the signals through the nerves to the brain and elsewhere in the body. Much has also been learned about how neural channels for information transmission become reinforced by use or may be replaced after damage; about which neurons or groups of neurons are committed to processing information derived from a particular organ or environmental location; and about many other issues concerning neural processes. But despite all this progress, neurobiology remains an infant discipline, at a stage of theoretical development comparable perhaps to that of genetics at the beginning of the twentieth century when Mendel's laws of heredity were rediscovered. Those things that count most remain shrouded in mystery: how physical phenomena become mental experiences (the feelings and sensations, called "qualia" by philosophers, that contribute the elements of consciousness) and how out of the diversity of these experiences emerges the mind, a reality with unitary properties such as free will and the awareness of self that persist throughout an individual's life.

I do not believe that the mysteries of the mind are unfathomable; rather, they are puzzles that humans can solve with the methods of science and illuminate with philosophical analysis and reflection. And I will place my bets that, over the next half century or so, many of these puzzles will be solved. We shall then be well on our way toward heeding the injunction "Know thyself."

Is evolution "just" a theory? In the next chapter I assert that evolution is indeed a theory. But it is a theory in the scientific sense, not just a guess or hunch but a well-integrated body of scientific knowledge supported by innumerable observations and experiments. Therefore, evolution is a fact as well as a theory.

Why Is Evolution a Theory?

EVOLUTION IS A THEORY and also a fact. This double claim may seem surprising, so it needs to be explained. Let me explain first why evolution is a theory and then why it is a fact.

When scientists talk about the "theory" of evolution, they use the word differently than people do in ordinary speech. In everyday speech, *theory* often means "guess" or "hunch" or "speculation." Somebody might say, "I have a theory about where Osama bin Laden is hiding." Or someone might express an idea such as "People who live in the suburbs are fatter than people who live in the city," and somebody else might respond, "That is your theory, but you have no proof that it is a fact." In science, however, the moniker *theory* is most properly used for a well-substantiated explanation of some aspect of the natural world that incorporates observations, facts, laws, inferences, and tested hypotheses. Although scientists sometimes use the word *theory* more casually, for tentative explanations that lack substantial supporting evidence, such tentative explanations are more accurately termed "hypotheses." It is important to distinguish these casual uses of the word *theory* from its use to describe notions such as evolution that are supported by overwhelming evidence.

Science has many other powerful theories besides evolution. The heliocentric theory says that the earth revolves around the sun rather than vice versa. The atomic theory says that all matter is made up of atoms. And the cell theory says that all living things consist of cells. According to the theory of evolution, organisms are related by descent from common ancestors. Species that share a recent common ancestor are more similar to each other than species whose last common ancestor is more remote. Humans and chimpanzees are, in configuration and genetic makeup, more similar to each other than they are to baboons, elephants, or kangaroos. There is a multiplicity of species because organisms gradually change from generation to generation, and different lineages change in different ways in response to different environments.

The word *fact*, like the word *theory*, has a different meaning in science than in common usage. A scientific fact is an observation that has been confirmed again and again, usually beyond reasonable doubt: that common salt is composed of chlorine and sodium or that DNA is made up of the four nucleotides represented as A, C, G, and T, for example. In common language *fact* is used more broadly, so as to include knowledge based on direct observation or experience, which often may not have been confirmed beyond reasonable doubt.

Scientists agree that the evolutionary origin of animals and plants is a scientific conclusion beyond reasonable doubt. They place it beside such established concepts as the roundness of the earth, its revolution around the sun, and the atomic composition of matter. That evolution has occurred is, in ordinary language, a fact, not just a theory.

An objection may now be raised. How can the factual claim

that evolution has occurred be asserted if no one has observed the evolution of species—for example, the evolution of humans and chimps—from a common ancestor, much less replicated it by experiment? Is it not true that science relies on observation, replication, and experimentation? This is indeed true, but what scientists observe and experiment with are not the concepts or general conclusions of theories but their consequences.

Copernicus's heliocentric theory affirms that the earth revolves around the sun. Scientists soon accepted this claim because of numerous confirmations of its predicted consequences, even though no one had yet observed the earth revolving around the sun. Even now, nobody has observed the annual revolution of the earth around the sun, not even astronauts. We accept that matter is made of a diversity of atoms, even though no one has seen the atoms, because of corroborating observations and experiments in physics and chemistry. Powerful microscopes have made it possible in recent years to examine materials at huge magnification, where the external configuration of presumptive atoms and molecules is observed. This is still a far cry from observing the atoms of current atomic theory, with their protons, electrons, neutrons, and other particles. Scientists may very well directly observe atoms and their detailed composition and configuration eventually, but the atomic theory of the composition of matter does not depend on such observation. Nor would direct observation of the existence of atoms contribute much to the atomic theory, which indeed encompasses much more knowledge than simply claiming that atoms exist.

The theory of evolution also depends on immensely numerous observations and experiments that confirm the consequences of the theory. For example, the claim that humans and chimpanzees are more closely related to each other than they are to baboons leads to the prediction that the DNA of humans and chimps is more similar than that of chimps and baboons. To test this prediction, scientists select a particular gene, examine its DNA structure in each species, and thus corroborate the inference. Experiments of this kind are replicated in a variety of ways to gain further confidence in the conclusion. And so it is for myriad predictions and inferences between all sorts of organisms.

The theory of evolution makes statements about three different but related issues: (1) the fact of evolution, that is, that organisms are related by common descent; (2) evolutionary history—the details of when lineages split from one another and of the changes that occurred in each lineage; and (3) the mechanisms or processes by which evolutionary change occurs.

The first issue is the most fundamental and the one established with utmost certainty. Darwin gathered much evidence in its support, but evidence has accumulated continuously ever since, derived from all biological disciplines. The evolutionary origin of organisms is today a scientific conclusion established beyond reasonable doubt, endowed with the kind of certainty that scientists attribute to well-established scientific theories in physics, astronomy, chemistry, and molecular biology. This degree of certainty beyond reasonable doubt is, as stated above, what is implied when biologists say that evolution is a "fact"; the evolutionary origin of organisms is accepted by virtually every biologist.

The theory of evolution goes far beyond the general affirmation that organisms evolve. The second and third issues—seeking to ascertain evolutionary history and to explain how and why evolution takes place—are matters of active scientific investigation. Some conclusions are well established. One, for example, is that chimpanzees are more closely related to humans than is either of these two species to baboons or to other monkeys. Another conclusion is that natural selection, the process postulated by Darwin, explains the configuration of such adaptive features as the human eye and the wings of birds. Many matters are less certain, others are conjectural, and still others—such as the characteristics of the first living things and the precise time when they came about—remain largely unknown, as discussed in chapter 5.

However, uncertainty about these issues does not cast doubt on the fact of evolution. Similarly, we do not know all the details about the configuration of the universe and the origin of the galaxies, but this is not a reason to doubt that the galaxies exist or to throw out all we have learned about their characteristics. Evolutionary biology is one of the most active fields of scientific research at present, and significant discoveries continually accumulate, supported in great part by advances in other biological disciplines.

The study of biological evolution has transformed our understanding of life in the world. Biological evolution accounts for three fundamental features of the world around us: the similarities among living beings, the diversity of life, and the adaptations of organisms—why animals have eyes for vision, wings for flying, and gills for breathing in water. Evolution accounts for the appearance of humans on Earth and reveals our relationship with other living things. Evolution is now the central organizing principle that biologists use to understand the world. As Theodosius Dobzhansky, one of the greatest evolutionists of the twentieth century, put it, "Nothing in biology makes sense except in the light of evolution."

Learning about evolution also has practical value. Modern biology has broken the genetic code, developed highly productive crops, and provided knowledge for improved health care. The theory of evolution has made important contributions to society. Evolution explains why many human pathogens have developed resistance to formerly effective drugs and suggests ways of confronting this increasingly serious health problem. Evolutionary biology has contributed importantly to agriculture by explaining the relationships between wild and domesticated plants and between animals and their natural enemies. An understanding of evolution is indispensable for establishing sustainable relationships with the natural environment.

One example of how evolutionary knowledge can help to solve an important world health emergency has been described in a document from the National Academy of Sciences and the Institute of Medicine:

In late 2002 several hundred people in China came down with a severe form of pneumonia caused by an unknown infectious agent. Dubbed "severe acute respiratory syndrome," or SARS, the disease soon spread to Vietnam, Hong Kong, and Canada and led to hundreds of deaths. In March 2003 a team of researchers at the University of California, San Francisco, received samples of a virus isolated from the tissues of a SARS patient. Using a new technology known as a DNA micro array, within 24 hours the researchers had identified the virus. . . . An understanding of evolution was essential in the identification of the SARS virus. . . . Furthermore, knowledge of the evolutionary history of the SARS virus gave scientists important information about the disease, such as how it is spread. (*Science, Evolution, and Creationism*, 2008, p. 5)

Darwin and other nineteenth-century biologists found compelling evidence for biological evolution in the comparative study of living organisms (anatomy), in their geographic distribution (biogeography), and in the fossil remains of extinct organisms (paleontology). Since Darwin's time, the evidence from these sources has become stronger and more comprehensive, while biological disciplines that emerged more recently—genetics, biochemistry, ecology, animal behavior (ethology), neurobiology, and especially molecular biology—have supplied powerful additional evidence and detailed confirmation.

Darwin surely would have been pleased by the enormous accumulation of paleontological evidence—for example, the discovery of fossils of organisms intermediate between major groups, such as *Archaeopteryx*, intermediate between reptiles (dinosaurs) and birds, and *Tiktaalik*, intermediate between fish and tetrapods, and the numerous fossils and diverse species of hominids intermediate between apes and *Homo sapiens*. But there are good reasons to believe that Darwin would have been most pleased and impressed with the overwhelming evidence for evolution and precise information about evolutionary history provided by molecular biology, a source of evidence and documentation of history that Darwin could not even have imagined.

Molecular biology, a discipline that emerged in the second half of the twentieth century, following the 1953 discovery of the double-helix structure of DNA, the hereditary material, now provides the strongest evidence yet of the evolution of organisms. In the next chapter, I explain DNA and how dramatically it reinforces the evidence for evolution.

Before I turn to those matters, I need to make an important observation germane to the topic of this chapter. In science, every hypothesis and theory remains forever subject to the possibility of rejection or replacement by a new theory. No matter how much evidence may have been accumulated in support of a theory, the possibility always remains that the reasons the observed phenomena and experimental results occur can derive from some other theory that accounts not only for the known facts but also for other as yet unexplained facts. Nevertheless, it seems unlikely that theories as extensively corroborated by contributions from so many disciplines as the evolution of organisms will ever be rejected or replaced. Surely, however, the theory of evolution will be further developed and advanced. Similarly, it seems utterly unlikely that the heliocentric theory or the molecular composition of matter will ever be rejected.

Let us now have a look at DNA, the "chemical of life," as it has been called, and at the evidence for evolution that comes from molecular biology.

3

What Is DNA?

D NA IS SHORT FOR deoxyribonucleic acid. The chemical structure of DNA is a double helix made up of two
complementary strands, which are long chains of four different kinds of nucleotides: adenine (A), cytosine (C), guanine
(G), and thymine (T). The genetic information embodied in
DNA accounts for three fundamental aspects of life: (1) all
living processes in organisms, (2) the precision of biological
heredity, and (3) biological evolution.

First, DNA holds the genetic information that directs all
life processes. The information is encased in long sequences
of the four nucleotides in a manner analogous to the way semantic information is conveyed by sequences of letters of the
English alphabet. The amount of genetic information in the
DNA of organisms is enormous because the total length of
the DNA molecules of an organism is huge. For example, the
human genome—that is, the DNA that each human inherits
from each parent—is 3 billion letters long. Printing one human genome would require 1,000 books, each 1,000 pages
long, with 3,000 letters (equivalent to about 500 words) per
page. Scientists do not print full genomes of humans or other

organisms; rather, the DNA information is stored electronically in computers.

A second attribute of DNA is that it accounts for the precision of biological heredity. The two strands in the DNA double helix are complementary, both carry the same genetic information, and either one of the two strands can serve as a template for the synthesis of a complementary strand identical to the original complementary strand. Each of the four nucleotides pairs in the complementary strand with only a particular one of the other three. A pairs only with T, and C pairs only with G. For example, if a short segment of one strand consists of the sequence ATTCAGCA, the complementary strand will be TAAGTCGT. This complementarity accounts for the fidelity of biological heredity. In the process of reproduction, the two helically coiled strands unwind and each serves as a template for the synthesis of a complementary strand so that the two daughter double helices are identical to each other and to the mother molecule. Thus the sequence ATTCAGCA would direct the synthesis of a complementary strand, TAAGTCGT, identical to its partner in the original DNA molecule. Similarly, the original strand TAAGTCGT will direct the synthesis of a complementary strand, ATTCAGCA, resulting in a double helix identical in sequence to the sister and to the mother double helix.

The third fundamental property of DNA is mutation, which makes possible the evolution of organisms. The information encoded in the nucleotide sequence of DNA is, as a rule, faithfully reproduced during replication, so that each replication results in two DNA molecules that are identical to each other and to the maternal molecule, as just explained. The fidelity of the process is enormous but not perfect. Occasionally mutations occur in the DNA molecule during replication so that daughter cells differ from the parental cells (and from each other) in the nucleotide sequence or in the length of the DNA. Mutations often involve a single letter (nucleotide), but occasionally mutations may encompass several or many letters. A mutation first appears in the DNA in a single cell of an organism, and the new, changed DNA is passed on to all cells descended from the first.

A notorious example of a mutation with important consequences in recent European history accounts for hemophilia, typically a fatal disease, which is determined by a mutation in the X chromosome. Gender in humans is determined by X chromosomes. Women have two X chromosomes; men have one X and one Y chromosome. Women with the hemophilia mutation in one X chromosome do not suffer from the disease but transmit it to those sons (half on average) who happen to inherit the X chromosome with the hemophilia mutation. A mutation for hemophilia occurred in one of Queen Victoria's X chromosomes. The mutation was transmitted through her daughters and granddaughters to the Russian, Spanish, and other royal families of Europe. Czarevitch Alexis, the only son of Czar Nicholas II of Russia, inherited hemophilia from his mother, Alexandra, Queen Victoria's granddaughter. Prince Alfonso, heir to the Spanish crown, inherited it from another granddaughter, Queen Ena, the wife of King Alfonso XIII. Political historians believe that the hemophilia of the throne's heirs contributed to the fall of the two royal families.

The mutations that count in evolution are those that occur in the sex cells (eggs and sperm) or in cells from which the sex cells derive, because these are the cells that produce the next generation. Mutations that happen in other cells often will be of little consequence and pass unnoticed. Some mutations may, however, give rise to cancer and other diseases. There also are notorious cases in which, for example, a person may have eyes of two different colors. Such instances are rare because mutations are rare.

Mutation rates have been measured in a great variety of organisms, mostly for mutants that exhibit conspicuous effects, such as changing eye color or metabolic effects that cause disease. In humans and other animals, rates of mutation typically range between one mutation for every hundred thousand cells and one mutation for every million cells. Although mutation rates are very low, new mutations appear every generation in every species because there are many individuals in each species and many genes in each individual. The human population consists of more than 6 billion people. If any given mutation occurs once in every million people, living humans would collectively carry six thousand newly occurred mutations of every possible mutation.

DNA mutations make possible the evolution of organisms. But if mutations were much more frequent than they are, they would cause multiple defects and even total disorganization. Mutations are more likely to cause defects or diseases than to be beneficial because they disrupt the established DNA sequence of an organism, which sequence has been selected over thousands of generations in a way that facilitates the survival and reproduction of an organism in the environments where it lives. But as explained above, the process of mutation provides each generation with many new genetic variations, which, so long as they are not lethal or greatly harmful, are added to the mutations carried over from previous generations. That is, species do not consist of genetically homogeneous individuals but rather of individuals that differ from each other by numerous mutations. Thus it is not surprising to see that when new environmental challenges arise, species are able to adapt to them. More than a hundred insect species, for example, have developed resistance to the pesticide DDT in parts of the world where spraying has been intense. Although these insects had never encountered this synthetic compound in their past evolutionary history, preexisting mutations allowed some of them to survive in the presence of DDT. That "adaptation" was rapidly multiplied by natural selection because only the DDT-resistant insects survived and produced the following generations.

The resistance of disease-causing bacteria and parasites to antibiotics and other drugs is a consequence of the same process. When an individual receives an antibiotic that kills the specific bacteria causing a disease—say, tuberculosis—the immense majority of the bacteria die, but one in several million may have a mutation that provides resistance to the antibiotic. These resistant bacteria will survive and multiply, and that antibiotic will no longer cure the disease in that individual or in any other person who becomes infected with the resistant bacteria. This is why, most recently, medicine treats bacterial diseases with cocktails of antibiotics. If the incidence of a mutation conferring resistance to a given antibiotic is one in a million, the incidence of one bacterium carrying three mutations, each conferring resistance to one of three antibiotics, is one in a quintillion (one in a million million million); it is not likely, if not altogether impossible, for such bacteria resistant to all three antibiotics to exist in any infected individual.

Mutations that make bacteria resistant to the antibiotics to which they are exposed are beneficial to the bacteria and are rapidly spread through the bacterial populations. Mutations beneficial to humans or to our ancestors have happened in our evolutionary history and account for features that are distinctively human. One example is *FOXP2*, the first gene ever identified with a function related to language. The *FOXP2* gene is shared by mammal species, where it plays an important role in the development of lung epithelium. Two mutations occurred in the human lineage that have allowed the *FOXP2* gene to acquire a new role in humans, namely in language, where it plays a crucial role. The mutated gene was thereby favored by natural selection in our ancestors and rapidly spread through human populations. This is thought to have occurred around 200,000 ya, which roughly coincides with the evolution of modern humans. Mutations have recently been identified in a human family where a dysfunctional *FOXP2* gene results in severe speech impairment.

A kind of mutation with very significant consequences in evolution is gene duplication. If one gene becomes duplicated, one copy may continue playing the original role, while the other may acquire a new function. One well-known example concerns the evolution of the globin genes and their association with respiration. An ancestral duplication allowed one gene to be involved in oxygen metabolism in muscle (myoglobin), the other, in blood (hemoglobin). Further duplications of the hemoglobin gene made possible the evolution of the very efficient modern tetramer hemoglobin, which consists of two components (polypeptides) of one kind and two of another kind. Hemoglobin A, which makes up 98 percent of human adult hemoglobin, consists of two alpha and two beta polypeptides. Additional duplications have resulted in additional specialization, such as the gamma gene that is active in the human fetus.

DNA has been called the "master molecule" because it directs the development and functioning of organisms. DNA in interaction with the environment (that is, the cell environment as well as the external environment, such as a river or a forest where an organism may live) determines the features of organisms, that is, their "phenotype," which (broadly understood) includes not only their appearance but also their behavior. DNA exercises its "mastery" through two intermediate steps. In the first step, the information in the sequence of nucleotides in the DNA double helix becomes "transcribed," copied into a new kind of molecule, called "messenger RNA" (mRNA; RNA stands for ribonucleic acid). In the process of transcription, one of the DNA strands is copied following the same rules as in the duplication of DNA except that RNA has uracil (U) rather than thymine. Thus the DNA segment ATTCAGCA becomes transcribed in RNA as UAAGUCGU.

The DNA is in the nucleus of the cell, and it is there that it is transcribed into messenger RNA. This mRNA now moves from the nucleus to the cytoplasm, the main body of the cell, where it is "translated" into a protein or polypeptide. Polypeptides make up proteins, and some proteins consist of more than one polypeptide. For example, the hemoglobin that prevails in our arteries and veins is made up of four polypeptides, as noted above, two of one kind (alpha hemoglobin) and two of another kind (beta hemoglobin). Proteins are chains of amino acids. There are twenty kinds of amino acids. The DNA information conveyed in the sequence of nucleotides becomes translated into a sequence of amino acids.

Remember now that proteins are made up of twenty different kinds of amino acids. How is the translation from the four-letter DNA and four-letter RNA language into the twenty-letter language of proteins accomplished? The trick is that the mRNA is read three letters at a time. Each set of three consecutive letters is called a "triplet" or a "codon." The triplet AGC, for example, is read as serine, and triplet AUG is read as methionine. Serine and methionine are two of the twenty amino acids. There are sixty-four possible triple combinations (codons) of four kinds of letters (nucleotides). So the genetic code is redundant; some amino acids are coded by several kinds of codons. For example, serine is coded by the triplet AGU in addition to AGC.

There are two types of proteins. Some are important structural components of organisms; for example, collagen is the main protein component of bone. Other proteins are enzymes, catalysts that mediate chemical reactions in all organisms. Enzymes may be seen as molecular machines that mediate all living processes inside cells; that is, enzymes catalyze the transformation of one substance into another. Enzymes are tremendously effective machines; they are thousands or millions of times more effective than the most effective human-made machines. The goal of so-called nanotechnology is precisely to design molecules that may act as enzymes and to use these enzymelike machines to yield desired products with an efficiency immensely greater than the machines now used in human industries, including information technologies. Most chemical products in cells are outcomes of a series of chain reactions, each catalyzed by a different enzyme. The genetic information that accounts for the makeup of proteins is in all cases embodied in the nucleotide sequence of the DNA, which is the molecule that accounts for the fidelity of biological heredity but also accounts for evolution as a consequence of DNA mutations.

Chapter 2 affirms that evolution is both a theory and a fact. In the next chapter we review the sorts of evidence that support that affirmation, using examples from paleontology, anatomy, biogeography, and molecular biology. The evidence available to the experts is enormous: hundreds of books and many thousands of research papers published every year in scores of scientific journals. As stated in the introduction, scientists accept the evolution of organisms with the same confidence as they accept other well-confirmed theories, like the atomic theory and the genetic theory of heredity.

An *Archaeopteryx* fossil showing traits intermediate between dinosaurs (reptiles) and birds. *Archaeopteryx* were small animals, about the size of a crow, that lived in the Late Jurassic period, about 60 Ma, in central Europe. Several well-conserved fossil specimens have been discovered in Bavaria, the first in 1861, the most recent in 2005. *Archaeopteryx* exhibits some features similar to small bipedal dinosaurs, including most of the skeleton. Other features, such as the skull, the beak, and the feathers, are birdlike, as clearly shown in the fossils. (Courtesy of the Museum für Naturkunde, Berlin)

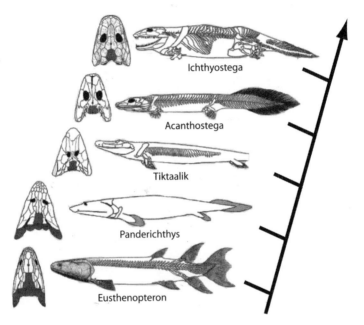

Ichthyostega

Acanthostega

Tiktaalik

Panderichthys

Eusthenopteron

Tiktaalik and other fossils intermediate between fish and amphibians (tetrapods), which lived between 385 Ma (*Eusthenopteron*, the most fishlike of those shown in the figure) and 359 Ma (*Ichthyostega*, with definite amphibian features). Several specimens of *Tiktaalik* were discovered in 2006 in Late Devonian river sediments, dated about 380 Ma, on Ellesmere Island in Nunavut, Arctic Canada. (Adapted from Per Ahlberg and Jennifer A. Clark, "Palaeontology: A Firm Step from Water to Land," *Nature* 440 [2006]: 747)

Normal						
Valine	Histidine	Leucine	Threonine	Proline	Glutamic acid	Glutamic acid
1	2	3	4	5	6	7

Sickle-cell anemia						
Valine	Histidine	Leucine	Threonine	Proline	Valine	Glutamic acid

The first seven amino acids of the beta chain of human hemoglobin. The beta polypeptide consists of 146 amino acids. A mutation in the gene coding for the beta polypeptide results in the substitution of the amino acid valine for glutamic acid. Individuals who have inherited the mutated gene from both parents cannot produce normal hemoglobin A and thus suffer from a severe disease known as sickle-cell anemia. Individuals who have inherited the normal gene from one parent but the mutated gene from the other parent do not suffer from the disease and, in addition, are protected against malaria, a malignant infection widespread in the tropics. This state of affairs is not unusual: people with two forms of a gene, one mutated but not the other, each inherited from one of the parents, are better off than individuals with two identical copies of the same gene. But the case of the beta hemoglobin gene is more extreme than most others.

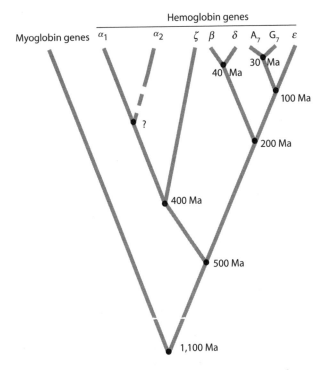

Myoglobin genes

Hemoglobin genes

α_1 α_2 ζ β δ A_γ G_γ ε

30 Ma

40 Ma

100 Ma

?

200 Ma

400 Ma

500 Ma

1,100 Ma

Evolutionary history of the globin genes, which account for oxygen metabolism. Several gene duplications, which occurred over long intervals of time, made it possible for each of the two duplicates to acquire somewhat different functions. The dots indicate points at which ancestral genes duplicated, giving rise to new gene lineages. The numbers next to the dots indicate (in millions of years) the times at which the duplications occurred. An early duplication allowed for one gene to be involved in oxygen metabolism in muscle (myoglobin), the other gene in oxygen transport (hemoglobin) in blood. Further duplications of the hemoglobin genes made possible additional specializations.

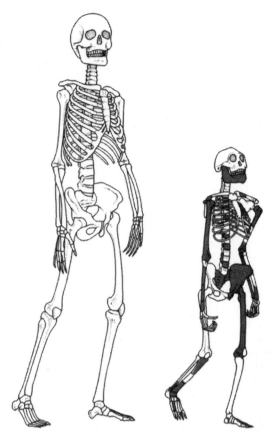

Skeleton of a modern human compared to Lucy, exemplar of *Australopithecus afarensis*, an ancestral species of modern humans that lived around 3.5 Ma and had bipedal gait but a small body and a small brain. About 40 percent of Lucy's skeleton (shaded in the figure) was found at a single site.

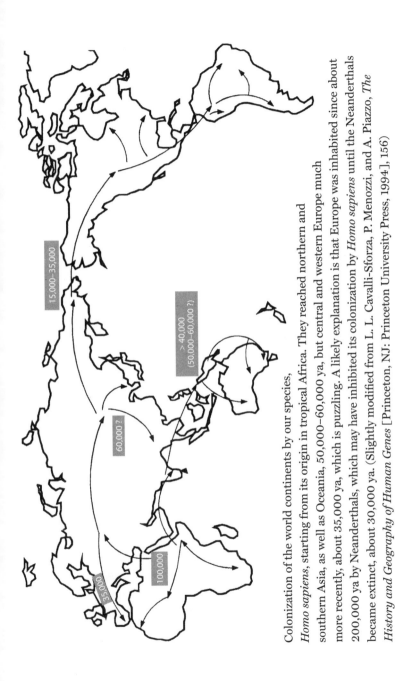

Colonization of the world continents by our species,

Homo sapiens, starting from its origin in tropical Africa. They reached northern and southern Asia, as well as Oceania, 50,000–60,000 ya, but central and western Europe much more recently, about 35,000 ya, which is puzzling. A likely explanation is that Europe was inhabited since about 200,000 ya by Neanderthals, which may have inhibited its colonization by *Homo sapiens* until the Neanderthals became extinct, about 30,000 ya. (Slightly modified from L. L. Cavalli-Sforza, P. Menozzi, and A. Piazzo, *The History and Geography of Human Genes* [Princeton, NJ: Princeton University Press, 1994], 156)

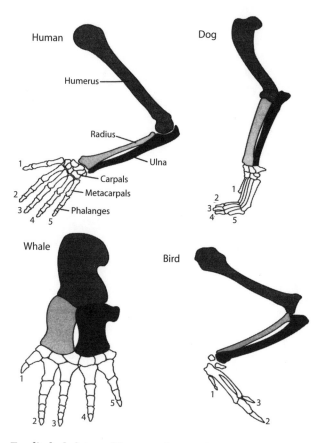

Forelimb skeleton of four vertebrates, showing similar and similarly arranged bones used for different functions in human, dog, whale, and bird. Evolutionists account for the striking similarity of structures serving disparate functions by common evolutionary origin with gradual modification as organisms adapt to their different ways of life. The similarity of structures is considerable among closely related organisms, different kinds of rodents for example, but becomes less so between species less closely related in their evolutionary history. Thus, morphological similarities are fewer between mammals and birds than they are among mammal species (a cat and a dog, for example), and they are still fewer between mammals and fishes. Morphological similarities not only are evidence of evolution but also help evolutionists reconstruct the evolutionary history of organisms.

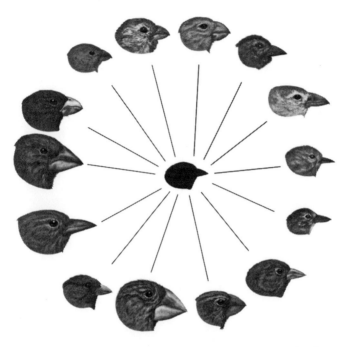

Darwin's finches. Fourteen species from the Galápagos Islands
that evolved from a common ancestor. Different species feed on
different foods and have evolved beaks adapted to their feeding
habits. The Galápagos Islands lie on the equator, six hundred miles
off the west coast of South America. They are, like the Hawaiian
Islands, of volcanic origin and formed at different times, starting
around 4.5 Ma. The Galápagos Islands were colonized by one finch
species from continental South America about 3 Ma. New finch
species evolved as they colonized available niches in the various
islands. (Adapted from "Evolution, The Theory of," courtesy of
Encyclopædia Britannica, Inc.)

Do All Scientists Accept Evolution?

THE OVERWHELMING MAJORITY of biologists accept evolution. Those who know professionally the evidence for evolution cannot deny it. Scientists agree that the evolutionary origin of animals and plants is a scientific conclusion beyond reasonable doubt. The evidence is compelling and all-encompassing because it comes from all biological disciplines including those that did not exist in Darwin's time. In the second half of the nineteenth century, Darwin and other biologists obtained convincing evidence from a variety of disciplines, which had reached early maturity during the nineteenth century: anatomy, embryology, biogeography, geology, and paleontology. Since Darwin's time, the evidence for evolution has become much stronger and more comprehensive, coming not only from traditional sources but also from recent disciplines such as genetics, biochemistry, ecology, ethology, neurobiology, and molecular biology.

As we saw in chapter 2, evolutionary research concerns three different subjects. Because the evidence is so overwhelming, the first subject—evidence for evolution—no longer engages the interest of biologists except when explaining evolution to the public or arguing with those who refuse to accept evolution. Although not sought and no longer needed, the evidence for the fact of evolution continues to accumulate.

Evolutionary research nowadays seeks greater knowledge of how evolution happens—the causes or mechanisms of the process—and how it actually happened, the phylogeny of organisms, their evolutionary history. Some assertions focus on the investigation of the historical process of evolution in particular groups of organisms because these organisms are of interest to the investigators or in order to solve questions with practical applications in agriculture, medicine, or industry. For example, experts have ascertained that cultivated corn derives from a Mexican wild plant known as teosinte.

When an organism dies, it is usually destroyed by bacteria and other organisms and by weathering processes. On rare occasions, some body parts—particularly hard ones such as teeth, bones, and shells—are preserved by being buried in mud or protected in some other way from predators, decomposition, and weather, and they may be preserved indefinitely as fossils within the rocks in which they are embedded. (Mud and other sediments may over time become limestone and other sorts of rocks.)

The fossil record is incomplete. Of the small proportion of organisms preserved as fossils, only a tiny fraction have been recovered and studied by paleontologists. Nevertheless, in many cases the succession of forms over time has been reconstructed in considerable detail. One example is the evolution of the horse, which can be traced from *Hyracotherium*, an animal the size of a dog having several toes on each foot and teeth appropriate for browsing (eating tender shoots, twigs, and leaves of trees and shrubs), which lived more than 50 Ma, to *Equus*, the much larger modern horse, one-toed and with teeth appropriate for grazing (eating growing herbage). A fairly detailed succession of intermediate, gradually changing horses has been discovered, going from *Hyracotherium* to *Equus*.

Particularly interesting are fossils intermediate between major groups of organisms, of which many are known. Two examples were cited in chapter 2: *Archaeopteryx* and *Tiktaalik*. *Archaeopteryx*, which lived 60 Ma, is intermediate between reptiles and birds. It had feathers, clearly shown in the fossils, while the skeleton is reptilelike, although the skull and beak are also like those of a bird. Several examples of *Archaeopteryx* have been discovered over the years, the first in 1861, two years after the publication of Darwin's *Origin of Species*, to which Darwin referred in later editions of the book. The most recent *Archaeopteryx* fossil, discovered just a few years ago, is now available for viewing in a private museum in Wisconsin.

Numerous intermediates between fish and tetrapods (amphibians) have been discovered over the years, but one of the most significant is *Tiktaalik*, which was described only recently, in 2006. Several specimens of *Tiktaalik* were found on Ellesmere Island in Nunavut, Arctic Canada, in 380-million-year-old sediments. These sediments were especially explored by scientists because that is the time when the earliest tetrapods were thought to have evolved.

The fossil intermediates of greatest interest to most people are those between humans and their ancestral primates. Hundreds of fossil hominids have been discovered over the last hundred years, as described in chapter 1.

The skeletons of turtles, birds, horses, bats, humans, and whales are strikingly similar, in spite of the different ways of life of these animals and the diversity of their environments. The correspondence, bone by bone, can easily be seen in the limbs. From a designing or engineering perspective, it seems incomprehensible that a turtle and a whale should swim, a dog run, a person write, and a bird or a bat fly with forelimbs built of the same bones organized in similar structures. An engineer could design better limbs for each purpose. But if we accept that all these animals inherited their skeletal structures from a common ancestor and became modified only as they adapted to different ways of life, the similarity of their structures makes sense.

Early support for the theory of evolution came also from embryology, the science that investigates the development of animals from fertilized egg to time of birth or hatching. Fishes, lizards, birds, and humans develop in ways that are remarkably similar during the early stages but become more and more differentiated as the embryos approach birth. The similarities persist longer between animals that are more closely related (say, humans and monkeys) than between those less closely related (e.g., humans and sharks). Human embryos have gill slits. These slits are found in the embryos of vertebrates that never breathe through gills because their ancestors were fish, in which these structures first evolved. Evolution explains why structures form during early development that will disappear before birth.

NUMBER OF SPECIES OF PLANT AND ANIMAL GROUPS NATIVE TO HAWAIIAN ISLANDS. The right column shows the percentage of species that naturally occur only in Hawaii. Many other species of plants and animals have recently been introduced by humans.

	Number of native species	Percent endemic
Ferns	168	65
Flowering plants	1,729	94
Snails	1,064	>99
Drosophila flies	510	100
Other insects	3,750	>99
Land mammals	0	0

Darwin saw evidence of evolution in the geographic distribution of plants and animals. He observed, for example, that different Galápagos Islands had different kinds of tortoises and finches, which in turn were different from those found in continental South America. There are about fifteen hundred known species of *Drosophila* fruit flies in the world; nearly one-third of them live nowhere except in the Hawaiian archipelago, although its land area is less than 5 percent the area of California. Also in Hawaii are more than a thousand species of snails not found anywhere else. On the other hand, there are no mammals native to Hawaii, and most kinds of insects are also lacking. These patterns of diversity are explained by evolution. The islands of Hawaii and the Galápagos are extremely isolated and have had few colonizers. Those species that did colonize the islands found many unoccupied ecological niches, local environments suited to sustaining them that they occupied by diversifying into a variety of species. Native mammals are nonexistent in both Hawaii and the Galápagos because of the improbability that living mammals would reach these remote archipelagos.

Molecular biology, a discipline that emerged in the second half of the twentieth century, nearly a hundred years after the publication of *The Origin of Species*, provides the strongest evidence of the evolution of organisms. Molecular biology proves evolution in two ways: first, by showing the unity of life in the nature of DNA and the workings of organisms at the level of enzymes and other protein molecules (as explained in chapter 3); second, by making it possible to reconstruct evolutionary relationships that were previously unknown and to confirm, refine, and time all evolutionary relationships among living organisms.

DNA and proteins have been called "informational macromolecules" because they are long linear molecules made up of sequences of units—nucleotides or amino acids—that embody evolutionary information in the sequence of their components, similar to how semantic information is embodied in sequences of letters of the alphabet. Comparing the sequence of the components in two macromolecules establishes how many units are different. Because evolution usually occurs by changing one unit at a time, the number of differences is an indication of the recency of common ancestry. Thus the inferences from paleontology, biogeography, comparative anatomy, and other disciplines that study evolutionary history can be tested in molecular studies of DNA and proteins of living organisms by examining the sequences of nucleotides and amino acids.

Molecular evolutionary investigations can be performed only with *living* species of organisms. The DNA and protein sequences of organisms that lived long ago are not preserved in a manner amenable to investigation with the methods of molecular biology currently available. The oldest fossils so far investigated with molecular methods—which include, for example, Egyptian mummies, mammoths, and some Neanderthal remains—are only a few thousand to about sixty thousand years old. Molecular biology can, nevertheless, reconstruct ancestral evolutionary relationships, such as determining which older species is ancestral to, say, a particular set of living species but not to others.

Molecular evolutionary studies have three notable advantages over comparative anatomy and other classical disciplines: *precision, universality,* and *multiplicity.* First, the information is readily quantifiable. The number of units that are different is easily established when the sequence of units is known for a given macromolecule in different organisms. It is simply a matter of aligning the nucleotides or amino acids of the different species and counting the differences. The second advantage, universality, is that comparisons can be made between very different sorts of organisms. There is very little that comparative anatomy or paleontology can say when organisms as diverse as yeasts, pine trees, and human beings are compared, but there are numerous DNA and protein sequences that can be compared in all three. The third advantage is multiplicity. Each organism possesses thousands of genes and proteins, every one of which reflects the same evolutionary history. If the investigation of one particular gene or protein does not satisfactorily resolve the evolutionary relationship of a set of species, additional genes and proteins can be investigated until the matter has been settled.

Molecular biology is useful to the study of evolution in other ways as well. The widely different rates of evolution of different sets of genes open up the opportunity for investigating different genes to achieve different degrees of resolution in the tree of evolution. Evolutionists rely on slowly evolving genes for reconstructing remote evolutionary events but examine increasingly faster-evolving genes for reconstructing the evolutionary history of more recently diverged organisms.

It is now possible to make an assertion that would have delighted Darwin and will surely startle those unfamiliar with molecular evolutionary studies: gaps of knowledge in the evolutionary history of living organisms no longer need to exist. Molecular biology has made it possible to reconstruct the universal tree of life, the continuity of succession from the original forms of life, ancestral to all living organisms, to every species now living on Earth. The main branches of the tree of life have been reconstructed. More details about more and more branches of the universal tree of life are published in scores of scientific articles every month. The virtually unlimited evolutionary information encoded in the DNA sequence of living organisms allows evolutionists to reconstruct all evolutionary relationships leading to present-day organisms with as much detail as desired. All that is needed is to invest the necessary resources (time and laboratory expenses) in order to have the answer to any query with as much precision as wanted.

Evolution occurs by natural selection in living organisms that reproduce and mutate. How did life get started? Are there living organisms on other planets? We explore these issues in the next chapter, pointing out what we know and what we don't know, which is a lot, about the origin of life.

How Did Life Begin?

ALMOST ALL BIOLOGISTS agree that life originated spontaneously by natural processes on our planet from the same chemicals of which living organisms consist today, such as carbon, nitrogen, oxygen, and hydrogen. Scientists also agree that all organisms living on Earth derive from a single original form of life. Does that mean that we know how life began? Not quite. Although there are some good ideas and good experiments about the origin of life, there is not yet general agreement about how life might have started. What we do know, because the evidence is overwhelming, is that all life on Earth has evolved from a single origin. We'll return to this question later; now let's examine where the difficulties are in demonstrating how life originated.

We start with the question, What is life? Two essential properties of life are heredity and metabolism. Cells reproduce by dividing, by making copies of themselves. The daughter cells must inherit the same components that make up the mother cell for there to be continuity of life. These components include the instructions about the chemical machinery of the cell: what chemicals to make and how they will operate. The instructions are themselves encased in chemicals. But the synthesis of the instruction chemicals requires the chemical machinery of the cell. We have a chicken-and-egg problem to get the process started. The instructions tell how the chemical machinery will operate, but the instructions cannot be synthesized without the chemical machinery. Think of a document to be copied: you need the photocopy machine to do the copying. Better yet, think of electronic computers: you need the hardware to carry out the instructions, and you need the software to provide the instructions. In life, the software includes the instructions for making the computer.

The information-carrying constituents are DNA and RNA molecules. DNA and RNA are strings made up of four kinds of chemical components, represented by the letters A, C, G, and T in DNA and by A, C, G, and U in RNA; that is, U replaces T in RNA. As described in chapter 3, the hereditary information is contained in sequences of the four letters, similar to letters of the English alphabet. The machinery that carries out the chemical reactions (called "metabolism") consists of enzymes, which are proteins able to catalyze chemical reactions with great accuracy and at speeds much faster than any human-made machine. Proteins consist of long strings of twenty different components, called amino acids.

A cell consists of thousands and thousands of components, including thousands of different kinds of enzymes, carrying out with tremendous efficiency and precision thousands of chemical reactions, many of them precisely determined sequences of reactions. Graphic representations of cellular processes show them as extremely complex networks, something like a graphic image of the whole transportation system in the United States might look. There are interstate and state thruways and highways, all sorts of additional roads, trails, driveways, and access roads; add oil wells, refineries, and gas stations and cars, trucks, motorcycles, and other vehicles, as well as the factories where they are made; and further still there are rivers, waterways, harbors, and all sorts of boats, as well as airports, airways, airplanes, and the factories where they are made. The extremely complex U.S. transportation network is surely no more complex than the network of chemical constituents and reactions that take place in a cell.

Consider now the question, How did the transportation network of the United States get started? We would probably think of simple foot trails, followed later by roads, paved or not, suitable for carts and wagons. But it would be difficult to ascertain where the first trails were fashioned, which towns or villages they connected, and other characteristics. The first trails in North America were made a few thousand years ago. In contrast, life originated on Earth several thousand Ma, so determining the origins of the trails of life is all the more difficult.

If we want to know how life first started, we need to identify the primitive constituents that made up the simplest forms of life. From the start we encounter, as mentioned previously, the chicken-and-egg problem. We need information molecules, such as DNA or RNA, which convey from mother cell to daughter cells the information about what enzymes are to be synthesized and how to synthesize them. But the DNA or RNA molecules need to be synthesized, and for that we need metabolism, the chemical machinery of living processes. Returning to the computer metaphor, in life the software has the information about how the computer is to be built, but the software (heredity molecules) cannot be read without the computer (metabolic machinery). Once you have computers in existence, there is no problem. The problem is how to get the first computer.

In 1953, Stanley Miller, a graduate student in chemistry at the University of Chicago, simulated in a tabletop glass apparatus the conditions that may have occurred on our planet soon after its birth. He included some inorganic chemicals such as ammonia, methane, and hydrogen gasses and added water vapor and electric discharges to simulate lighting. After one week, amino acids and other compounds, such as urea, naturally found only in organisms, had formed in the five-liter glass flask where the experiment was conducted. Miller had thus demonstrated that organic compounds could be formed without the mediation of enzymes. Later experiments, under conditions more similar to those on primitive Earth, have confirmed that simple organic compounds can be formed spontaneously. This possibility is now taken for granted because of a multitude of experiments, but also because simple organic molecules have been found in meteorites falling on our planet, in comets, and even in interstellar gas clouds.

The question remains how these basic building blocks joined to form more complex molecules, such as enzymes and DNA, and living cells. One favored scenario is that once the earth cooled enough to allow oceans to form, something like the processes observed by Miller and others resulted in a broth of organic molecules (a "primordial soup"), which given enough time (many million years were available!), would produce, by chance combination of molecules, some replicating entity that would have evolved into life as we know it.

The chicken-and-egg problem still remains: How do we get the continuity of life, the hereditary molecules that specify the synthesis of the enzymes to carry out the life processes, before we have enzymes? An important advance occurred in the early 1980s when Thomas R. Cech and Sydney Altman independently discovered that some RNA molecules can catalyze chemical reactions, including their own synthesis, a discovery for which they received the Nobel Prize in 1989. This discovery contributed importantly to solving the chicken-and-egg problem, in that these RNA molecules, dubbed ribozymes, can fill the two roles of heredity and metabolism, roles mostly played by separate molecules, DNA and proteins, respectively, in current living organisms. Many scientists now believe that life went through an RNA-dominated phase, called the "RNA world," which preceded the current DNA world, where biological heredity is prevailingly encased in DNA molecules.

Recent efforts have addressed the issue of how ribozyme RNA molecules may have formed spontaneously on primordial Earth, eventually leading to the RNA world. Ribozymes, like other RNA molecules, consist of four kinds of nucleotides, represented by the letters A, C, G, and U, as pointed out earlier, and are made up of a limited number of nucleotides, say two dozen or so. But nucleotides are far from simple, made up of three molecular components: a ribose sugar, a phosphate, and a nitrogen base. The base is the only constituent that varies from one nucleotide to another. There are four different kinds of nitrogen bases, which correspond to the A, C, G, and U nucleotides. Scientists have recently shown how the ribose sugar may become spontaneously linked to the nitrogen bases C and U, which was the step most difficult to account for.

Once RNA molecules were formed that could reproduce by copying themselves but were subject to some error (mutation) in the synthesis of new RNA molecules, natural selection would occur, leading to greater molecular complexity and eventually to cells, first simple cells such as occur in bacteria, and later to more advanced ones, as in animals, plants, and other eukaryotic organisms. Natural selection is the differential reproduction of alternative hereditary variants. Once there were primitive cells able to reproduce, some would probably reproduce more effectively than others. The characteristics of the cells that reproduced more effectively would increase in frequency at the expense of those that reproduced less effectively. It stands to reason that cells that reproduced more effectively would often be those that had more precise heredity and more efficient metabolism.

So do we now know how life began? We don't. What we do know is that spontaneous chemical processes, in the absence of previous life and under conditions that plausibly may have existed on the early Earth, can give rise to organic compounds including those that are the fundamental building blocks of life, the nucleic acids that encase heredity and the enzymes that account for metabolism (i.e., the proteins that catalyze the chemical reactions that make up all living processes). There is no reason to assume that the intervention of a supernatural agent was required in order to account for the origin of life. Returning to the transportation network analogy, we now know that early foot trails could be made and how more advanced roads might be developed from them, but we don't know where the first trails were made or precisely how they were made. We also know that life originated on Earth only once, or if it originated more than once, all forms of life but one became extinct.

The reason we know that all living organisms derive from a single origin is that organisms now living on Earth share fundamental life processes that could only have emerged from a single origin. All organisms, from bacteria and other microscopic organisms to animals, plants, and fungi, share certain features. The list of features that could possibly take alternative configurations and yet are uniform throughout all life is very long. We can start the list with DNA, the same heredity molecule in all organisms, always made up of the same four nucleotides, although the chemical possibilities for other nucleotides are many. The thousands of enormously diverse proteins that exist in each organism are synthesized from different combinations, in sequences of variable length, of twenty amino acids, the same twenty in all proteins and in all organisms. Yet several hundred other amino acids exist in nature. The complex machinery by which the hereditary information is conveyed from the nucleus to the main body of the cell is everywhere the same: the sequence of nucleotides in the DNA is transcribed into a complementary sequence of RNA (dubbed "messenger RNA"), which becomes "translated" into specified sequences of amino acids that make up the proteins and enzymes that carry out all life processes. This translation involves specific RNA molecules ("transfer RNA") and RNA-protein complexes ("ribosomes") that are universally shared. The genetic dictionary that gives the translation from DNA sequence into amino acid sequence is also universally shared.

The universally shared properties of life could be extended. The unity of life reveals the genetic continuity and common ancestry of all organisms.

The earth is probably the only planet in the solar system that currently has life. There are about 100 billion stars in our galaxy, and many of them have planetary systems. And there are more than 100 billion galaxies in the universe. It may very well be that life exists elsewhere in the universe. Some planets, perhaps very many of them throughout the universe, may have life if temperature, chemical composition, and other features favorable for life occur, as seems possible given the immense number of galaxies, stars, and planets. Life occurs on Earth because conditions favorable for life exist on our planet. Given similar conditions and eons of time, life would probably come to exist on other planets.

Life as it may exist elsewhere would have features different from those discussed above that show the unique common origin of life on Earth. Even the basic chemical elements might be different; for example, silicon rather than carbon may combine with hydrogen, oxygen, and other elements to make up the basic molecules of life. In a few decades, or in a few centuries or millennia, our descendants may discover life elsewhere in the universe. Surely they will then also learn more than we now know about how the distinctive features of life on Earth came about.

I will next turn to a topic that is not strictly scientific but is of great interest to many people of faith and to philosophers, sociologists, and historians: the relationship between evolution and religious faith (or, more generally, between science and religion). Are they compatible or antagonistic?

6

Can One Believe in Evolution *and* God?

I AM CONVINCED that evolution and religious beliefs need not be in contradiction. Indeed, if science and religion are properly understood, they *cannot* be in contradiction because they concern different matters. Science and religion are like two different windows for looking at the world. The two windows look at the same world, but they show different aspects of that world. Science concerns the processes that account for the natural world: how planets move, the composition of matter and the atmosphere, the origin and adaptations of organisms. Religion concerns the meaning and purpose of the world and of human life, the proper relation of people to the Creator and to each other, the moral values that inspire and govern people's lives. Apparent contradictions only emerge when either science or belief, or often both, cross over their boundaries and wrongfully encroach upon one another's subject matter.

Science is a way of knowing, but it is not the only way. Knowledge also derives from other sources. Common experience, imaginative literature, art, and history provide valid knowledge about the world, and so do revelation and religion for people of faith. The significance of the world and human life, as well as matters concerning moral or religious values, transcend science. Yet these matters are important; for most of us, they are at least as important as scientific knowledge per se.

The proper relationship between science and religion can be, for people of faith, mutually motivating and inspiring. Science may inspire religious beliefs and religious behavior, as we respond with awe to the immensity of the universe, the glorious diversity and wondrous adaptations of organisms, and the marvels of the human brain and the human mind. Religion promotes reverence for the creation, for humankind as well as for the world of life and the environment. For scientists and others, religion is often a motivating force and source of inspiration for investigating the marvelous world of the creation and solving the puzzles with which it confronts us.

To some Christians, the theory of evolution seems incompatible with their religious beliefs because it is inconsistent with the Bible's narrative of creation. The first chapters of the biblical book of Genesis describe God's creation of the world, plants, animals, and human beings. A literal interpretation of Genesis seems incompatible with the gradual evolution of humans and other organisms by natural processes.

Even in the nineteenth century, shortly after Darwin's publication of *The Origin of Species*, some Christian theologians saw a solution to the apparent contradiction between evolution and creation in the argument that God operates through intermediate causes. The origin and motion of the planets could be explained by the law of gravity and other natural processes without denying God's creation and providence. Similarly, evolution could be seen as the natural process through which God brought living beings into the existence and developed them according to his plan. A. H. Strong, the president of Rochester Theological Seminary in New York State, wrote in his 1885 *Systematic Theology*: "We grant the principle of evolution, but we regard it as only the method of divine intelligence." He explains that the brutish ancestry of human beings is not incompatible with their exalted status as creatures in the image of God.

Gradually, well into the twentieth century, evolution came to be accepted by a majority of Christian writers. Pope Pius XII in his 1950 encyclical *Human Generis* (On the Human Race) asserted that biological evolution was compatible with the Christian faith. Pope John Paul II, in an address to the Pontifical Academy of Sciences on October 22, 1996, said: "New scientific knowledge has led us to realize that the theory of evolution is no longer a mere hypothesis. It is indeed remarkable that this theory has been progressively accepted by researchers, following a series of discoveries in various fields of knowledge. The convergence, neither sought nor fabricated, of the results of work that was conducted independently is in itself a significant argument in favor of this theory."

The compatibility of evolution with the Christian faith has been asserted by other mainstream Christian denominations. The General Assembly of the United Presbyterian Church in 1982 adopted a resolution stating that "Biblical scholars and theological schools . . . find that the scientific theory of evolution does not conflict with their interpretation of the origins of life found in Biblical literature." The Lutheran World Federation in 1965 affirmed that "evolution's assumptions are as much around us as the air we breathe and no more escapable. . . . [B]oth science and religion are here to stay, and . . . need to remain in a healthful tension of respect toward one another."

Similar statements have been advanced by Jewish authorities and leaders of other major religions. In 1984, the Ninety-fifth Annual Convention of the Central Conference of American Rabbis adopted a resolution stating: "Whereas the principles and concepts of biological evolution are basic to understanding science . . . we call upon science teachers and local school authorities in all states to demand quality textbooks that are based on modern, scientific knowledge and that exclude 'scientific' creationism."

The "Clergy Letter Project," signed by more than twelve thousand U.S. Christian clergy members, makes a similar point: "We the undersigned, Christian clergy from many different traditions, believe that the timeless truths of the Bible and the discoveries of modern science may comfortably coexist. We believe that the theory of evolution is a foundational scientific truth, one that has stood up to rigorous scrutiny and upon which much of human knowledge and achievement rests. To reject this truth or to treat it as 'one theory among others' is to deliberately embrace scientific ignorance and transmit such ignorance to our children. We ask that science remain science and that religion remain religion, two very different, but complementary, forms of truth."

One difficulty with attributing the design of organisms to the Creator is that imperfections and defects pervade the living world. Consider the human eye. The visual nerve fibers in the eye converge to form the optic nerve, which crosses the retina (in order to reach the brain) and thus creates a blind spot, a minor imperfection but an imperfection of design, nevertheless; squids and octopuses do not have this defect. Did the Creator have greater love for squids than for humans and thus exhibit greater care in designing their eyes than ours? Consider now the human jaw. We have too many teeth for the jaw's size, so that wisdom teeth need to be removed and orthodontists can make a decent living straightening the others. Would we want to blame God for this blunder? A human engineer would have done better. Evolution gives a good account of these imperfections.

Examples of deficiencies and dysfunctions in all sorts of organisms can be endlessly multiplied. The world of organisms also abounds in characteristics that, as in the behavior of predators killing and devouring their prey, might be characterized as "cruelties," an apposite qualifier if the cruel behaviors were designed outcomes of a being holding on to human or higher standards of morality. But the cruelties are only metaphorical when applied to the outcomes of natural selection or to animals, because these lack moral status.

Religious scholars in the past struggled with imperfection, dysfunction, and cruelty in the living world, which are difficult to explain if they are the outcome of God's design. The Scottish philosopher David Hume (1711–1776) set the problem succinctly with brutal directness: "Is he [God] willing to prevent evil, but not able? Then he is impotent. Is he able, but not willing? Then he is malevolent. Is he both able and willing? Whence then evil?" Evolution came to the rescue. As the theologian Aubrey Moore put it in 1891, "Darwinism appeared, and, under the guise of a foe, did the work of a friend." The theory of evolution, which at first seemed to remove the need for God in the world, now has convincingly removed the need to explain the world's imperfections as failed outcomes of God's design.

If we claim that organisms and their parts have been specifically designed by God, we have to account for the incompetent design of the human jaw, the narrowness of the birth canal, and our poorly designed backbone, less than fittingly suited for walking upright. People of faith would do well to acknowledge Darwin's revolution and accept natural selection as the process that accounts for the design of organisms, as well as for the dysfunctions, oddities, and cruelties that pervade the world of life. Evolution makes it possible to attribute these mishaps to natural processes (which have no moral implications) rather than to the direct creation or specific design of the Creator.

Evolution may contribute to a possible theological explanation of defects, dysfunctions, cannibalism, parasitism, predation, and other "evils" of the living world. Some antireligious authors, as well as other critics, have argued that the process of evolution by natural selection does not discharge God's responsibility for the dysfunctions and cruelties of the living world because, for people of faith, God is the Creator of the universe and thus would be accountable for its consequences, direct or indirect, immediate or mediated. If God is omnipotent, the argument would say, He could have created a world where such things as cruelty, parasitism, and human miscarriages would not occur.

One possible answer is to claim that God's deeds are inscrutable and humans are not entitled to seek understanding of God's purposes, much less to bring His actions into account.

This answer may seem to many unsatisfactory, because it simply evades the question instead of answering it. A different possible explanation is the following. Consider, first, human beings, who perpetrate all sorts of misdeeds and sins, even perjury, adultery, and murder. People of faith believe that each human being is a creation of God, but this does not imply that God is responsible for human crimes and misdemeanors. Sin is a consequence of free will; the flip side of sin is virtue. Christian theologians have expounded that if humans are to enter into a genuinely personal relationship with their maker, they must first experience some degree of freedom. The eternal reward of heaven calls for a virtuous life. The critics might say that this account does not excuse God, because God could have created humans without free will (whatever these "humans" may have been called and been like). But one could reasonably argue that "humans" without free will would be a very different kind of creature, a being much less interesting and creative than humans are. Robots are not a good replacement for humans; robots do not perform virtuous deeds.

Before modern physical science came about, God, according to some religious views, caused rain, drought, earthquakes, and volcanic eruptions to reward or punish people. This view implies that God caused the tsunami that killed 200,000 Indonesians a few years ago. That would seem incompatible with a benevolent God. However, we now know that tsunamis and other natural catastrophes come about by natural processes. Natural processes don't entail moral values. Critics might object that God could have created a different world, without catastrophes. Yes, according to some belief systems, God could have created a different world. But that would not be a creative universe, where galaxies form, stars and planetary systems come about, and continents drift. The world that we have is creative and more exciting than a static world.

Turn now to badly designed human jaws, parasites that kill millions of children, and a poorly designed human reproductive system that accounts for millions of miscarriages every year in the world. If these dreadful happenings come about by direct design by God, God would seem responsible for the consequences. If engineers design cars that explode when you turn on the ignition key, they are accountable. But if the dreadful happenings come about by evolution or other natural processes, there are no moral implications, because natural processes don't entail moral values. Some might object, once again, that God is ultimately responsible because God could have created a world without cruelties, parasites, or dysfunctionalities. But a world of life with evolution is much more exciting; it is a creative world where new species arise, complex ecosystems come about, and humans have evolved. This account will not satisfy some people of faith, and many unbelievers will surely find it less than cogent. But I am suggesting that it may provide the beginning of an explanation for many people of faith.

Some Christians oppose acceptance of evolution, human evolution in particular, because they hold to a literal interpretation of the Bible. According to the Statement of Belief of the Creation Research Society, founded in 1963, "The Bible is the Written Word of God, and because it is inspired throughout, all of its assertions are historically and scientifically true in the original autographs. To the student of nature this means that the account of origins in Genesis is a factual presentation of simple historical truths."

Many Bible scholars and theologians have long rejected a literal interpretation of the Bible as untenable, however, because the Bible contains mutually incompatible statements. The very beginning of the book of Genesis presents two different creation narratives. Extending through Chapter 1 and the first verses of Chapter 2 is the familiar six-day narrative, in which God creates human beings—both "male and female"—in his own image on the sixth day, after creating light, earth, firmament, fish, fowl, and cattle. In verse 4 of Chapter 2, a different narrative starts, in which God creates a male human, then plants a garden and creates the animals, and only then proceeds to take a rib from the man to make a woman.

Which one of the two narratives is correct and which one is in error? Neither one contradicts the other, I would say, if we understand the two narratives as conveying the same message, that the world was created by God and that humans are His creatures. But it seems to me that both narratives cannot be "historically and scientifically true" as postulated in the Statement of Belief of the Creation Research Society.

There are numerous inconsistencies and contradictions in different parts of the Bible, in the description of the return from Egypt to the promised land by the chosen people of Israel, for example, not to mention erroneous factual statements about the sun's circling around the earth and the like. Biblical scholars point out that the Bible is inerrant with respect to religious truth, not in matters that are of no significance to salvation. Thus, Augustine (354–430), one of the greatest Christian theologians of all time, wrote in his *De Genesi ad lit-*

teram (Literal Commentary on Genesis): "It is also frequently asked what our belief must be about the form and shape of heaven, according to Sacred Scripture. . . . Such subjects are of no profit for those who seek beatitude. And what is worse, they take up very precious time that ought to be given to what is spiritually beneficial to salvation." This was a point made already by other early fathers of the Church. Augustine added: "In the matter of the shape of heaven, the sacred writers did not wish to teach men facts that could be of no avail for their salvation." That is, the book of Genesis is not an elementary book of astronomy. Augustine also noted that in the Genesis narrative of creation, God creates light on the first day but did not create the sun until the fourth day. Augustine concluded that "light" and "days" in Genesis make no literal sense. The Bible is about religion, and it is not the purpose of the Bible's religious authors to settle scientific questions.

Similar statements from religious authorities, including some I cited earlier, have continued to the present time. In 1981, John Paul II asserted that the Bible itself "speaks to us of the origins of the universe and its makeup, not in order to provide us with a scientific treatise but in order to state the correct relationships of man with God and with the universe. Sacred Scripture wishes simply to declare that the world was created by God. . . . Any other teaching about the origin and makeup of the universe is alien to the intentions of the Bible, which does not wish to teach how heaven was made but how one goes to heaven."

In conclusion, my answer is:

"Yes, one can believe in both evolution and God." Evolution is a well-corroborated scientific theory. Christians and other people of faith need not see evolution as a threat to their beliefs. Many theologians and other people of faith see evolution as the process by which God creates the wonderful diversity of the living world. Thus I would add, paraphrasing theologian Aubrey Moore, that evolution is not the enemy of religion but, rather, its friend.

Bibliographic Note

The analogy in chapter 5 between the U.S. transportation system and the metabolic pathways of a cell is borrowed from James Trefil, Harold J. Morowitz, and Eric Smith, "The Origin of Life," *American Scientist* 97 (2009): 206–213. For recent evidence of RNA molecules as catalysts (ribozymes), see Saba Valadkhan, Afshin Mohammadi, Yasaman Jaladat, and Sarah Geisler, "Protein-Free Small Nuclear RNAs Catalyze a Two-Step Splicing Reaction," *Proceedings of the National Academy of Sciences* 106 (2009): 11901–11906; and Samuel E. Butcher, "The Spliceosome as Ribozyme Hypothesis Takes a Second Step," *Proceedings of the National Academy of Sciences* 106 (2009): 12211–12212. For experiments showing how information-coding nucleic acids may have arisen spontaneously, see Jack W. Szostak, "Systems Chemistry on Early Earth," *Nature* 459 (2009): 171–172; and Matthew W. Powner, Béatrice Gerland, and John D. Sutherland, "Synthesis of Activated Pyrimidine Ribonucleotides in Prebiotically Plausible Conditions," *Nature* 459 (2009): 239–242. For additional information about the topics in chapters 1–4 and 6, see Francisco J. Ayala, *Darwin's Gift to Science and Religion* (Joseph Henry Press, 2007). An extended in-depth treatment of human origins can be found in Camilo J. Cela-Conde and Francisco J. Ayala, *Human Evolution: Trails from the Past* (Oxford University Press, 2007).